The SEO Solution To Rank On The First Page Of Google For Free

By Allen Spindel, COO

Janis Spindel Serious Matchmaking, Inc.

December 2012

I0469281

Janis Spindel Serious Matchmaking, Inc. All rights reserved. You may not distribute this book in any way. You may not sell it, or reprint any part of it without written consent from the author, except for the inclusion of brief quotations in a review. Copyright 2012.

Table Of Contents

Chapter 1   Article Writing
                     Article Marketing Outline
                     Bonus 2,304 Tweets

Chapter 2   Word Press and Blogger Blogs

Chapter 3   Blog Talk Radio

Chapter 4   Press Releases

Chapter 5   DMOZ

Chapter 6   Squiddo Lenses

Chapter 7   Web Page Format

Chapter 8   Proprietary Blog-O-Sphere

HELLO AND WELCOME to the worlds most unique and effective SEO lesson plans. The SEO lessons you are about to study and implement are taken from real lessons I learned while building my site http://www.janisspindelmatchmaker.com and now will pass onto you. I don't need to charge $150-$500 an hour, as I get 25% of all the revenue I bring in from the online businesses I work with. You see, I put my money where my mouth is. In addition, I am a retired teacher living on a pension and really am doing this to keep my mind active. So enjoy, sit back, as I can assure you that if you follow my step-by-step process, you will see dramatic improvement in your search engine ranking. Please excuse any grammar errors as I am focusing only on the SEO content and not trying to impress you with big words. Having said that, lets begin. If you enter **Matchmakers** keyword Google search you will see my site comes up 3rd in the organic search, below the paid search, which is indicated in the yellow box. If you follow my lesson plans, you won't need to pay to be listed. This book is short, sweet and highly effective if you use it.

The day I retired from teaching physical education at Washington Irving High School in New York City, I began studying SEO to improve Janis Spindel's site. Each morning as my daughter went to school and Janis went to work, I too went to work by drinking green tea at various Starbucks. I would sit at a table for 2 1/2 hours at a time and study. When I say study, I mean study. All I had was an old school Blackberry and a marble notebook and would learn from various experts. The major influential people that helped shape this course are Jeff Herring who calls himself The Article Marketing Guy, Alex Mandossian from Teleseminar Secrets, Gene Plotkin from Creation of Success and 13 key books from Barnes and Noble that are part of my Library.

What I found out and what I want to convey to you so you don't waste your time or money on this course is there are no short cuts; there are no secret formulas, no tricks or illegal techniques. I have only learned and practice SEO skills that Google LOVES. I can't stress enough that if you impress Google with doing the pure old school, boring, mundane and methodical step by step lessons I will show you, there is no way your site won't rise in ranking. When I first began working on http://www.janisspindelmatchmaker.com in 2008, that site ranked on page 7 of a Google search for 7 industry keywords. If you enter **Matchmaker** keyword Google search you will see my site comes up 4th in the organic search.

As of this SEO course written in December 2012, my site ranks on the first page of Google without spending a dime. Let me repeat that while I flex my muscles and pat myself on the back. Janis Spindel never spent any money to advertise her upscale matchmaking service and therefore I must follow her 20 years of success rule.

Article Writing

Article writing is the most effective free online marketing tool ever created. Let me repeat that. Article writing is the best free online marketing tool ever created. I used E-zine Articles when I first began and still use them today.

They have very strict rules and if you follow them exactly as I outline, these highly authoritative sites can help drive your site higher than any organic SEO programing by placing permanent back links on the internet which Google will eat up. Go to E-zine Articles Guidelines you can learn what the rules are

Here is my take, my experience and by the way, I searched on E-zines the key word "article writing" at the time I began and Jeff Herring, the Article Marketing Guy came up. He had 77 articles about article marketing and all I did was read every single article, take notes and compiled my own article writing course to help www.janisspindelmatchmaker.com web site increase in ranking. It took me 8 weeks to plow through those articles while working at Starbucks.

From my own article marketing course I wrote thanks to Jeff, I was able to crank out 40 articles in no time and within a few weeks, noticed that my web site rose in ranking. **Matchmaking** keyword Google search screen shot below. Notice that my site ranked 4th on the first page of this organic search.

Let me begin from the end result, showing you the power of article writing. Again, as of the writing of this eBook, I wrote 72 total articles on e-zines, with 18,683 views and a 3.52% click rate. That means that 658 visitors actually had an interest in the subject matter enough to click on the links that writers can post at the end of each article. Industry standard for a click rate is 1.5%, so we are doing something right and have developed a great niche market.

If you follow the math, we had 652 (rounded down) clicks to our web site. If 10% of those web site visitors joined our level 1 membership and paid $25, (65 members) I just made 65 X $25 = $1,630 while I sleep. Remember the term while I sleep.

If 25% of those web site visitors joined our level 1 membership and paid $25, (163 members) I just made 163 X $25 = $4,075 while I sleep.

Following along the next level of membership, level 2, we have been averaging 20% of level 1 members joining level 2 and paying an additional $250. If my math is correct, that's 163 level 1 members converting to level 2, which are 32 members paying $250, which is an additional $8,150.

Add the level 1 member revenue of $4,075 plus level 2 member revenue of $8,150, and now I just made a total of over $12,000 while I sleep. There I go again with the term while I sleep. You see, once I wrote those articles, they stay on e-zines forever and work for me full time without food or water or asking for a raise.

Let's talk about the concept of article marketing.

When writing articles about your business, your hobby, your passion, you want to be able to "work once and get paid forever." This is the mindset you must have. Please understand this, write it down, follow it and live it. If you do, it's like having an army of workers who work day and night for you. So, here is how it goes.

Since I did this for my web site www.janisspindelmatchmaker.com, I wrote my first article titled *"Matchmaker Reveals 7 Steps on How to Find a Successful Marriage Minded Partner."*

It will take you only 30-minutes to write each article when you follow the steps I will outline. Here is the end result from just one 400-600 word article:

Chop it up into 3 blogs.

Chop each blog into 3 Tweets.

Chop each blog into 3 Facebook Posts.

Chop each blog into 3 LinkedIn Posts.

Chop each blog into 3 YouTube branded 2-minute videos.

Chop each blog into 3 Blog Talk Radio show titles.

Chop each blog into 3 Squiddo Lenses.

Take that 1 e-zine article, tweak it and create a free press release on I-Newswire.

Now, if my math is correct (I got a 94 on my graduate statistic course) that one 400-600 word article that took you 30-minutes to write just generated:

3 blogs.

9 tweets.

9 Facebook posts.

9 LinkedIn posts.

9 YouTube 2-minute videos.

9 Blog Talk Radio show titles.

9 Squiddo Lenses.

1 Free Press Release.

Article Marketing Outline

Here are the main points when learning and writing articles to drive organic niche traffic to your web site and or blog.

*E-Zine guidelines.
*2 links in resource box. One blog and one main web site if different.
*No self-promotion.
*Keyword title.
*Keyword summary.
*Keyword first paragraph.
*Body must match title.
*Must be 400-600 words per article.
*Must be informative.
*No third party reference, must be based 0n your personal expertise.
*Reader must walk away with information they can feel good about and use right away.
*Must provide value to earn the readers trust.
*Must be URRC - unique, remarkable, relevant, content.
*You must exclusively own content.
*Show tips, strategy, techniques, case study and analysis.
*Must use proper English, spelling, grammar, etc.
*Use long tail keywords phrases.
*Not written as a press release or sales pitch.
*Make article summary as a teaser for what reader can look forward to.
*Best title has "6 Ways To" rather than "Six Ways To."
*Plant a question in the readers mind.
*Use the "&" in the title only and use "and" in the body.
*The summary should draw the reader in by promising something in the article they want.

Now comes the good stuff.

"16 Ways To Crank out Thousands of Blogs From One Article"

Now, if that is of interest to you, read on.

Take a concern you have, take an area of interest, take an expertise or passion you have for something. Let's say its golf. My 89-year-old dad loves golf and has been playing for 60 years. He wants to make a brand about golf for senior citizens. So I am going to show you how he would go about this.

Write down 16 THINGS that relate to golf. I will use only 3 as an example so you get the idea.

1. Time to play golf.
2. Clothing to wear while playing golf.
3. Sun screen to use while playing golf.

Now for each of the 16 THINGS, (the 3 THINGS EXAMPLES ABOVE) write down 4 THINGS OF EACH THING. An example would be:

THING 1. Time to play golf.
1. Golf in the morning or afternoon
2. Golf at a public or private course.
3. Golf in the winter vs. summer golf.
4. Golf for business or recreation.

So for the one THING: Time to play golf, we developed 4 THINGS OF THE THING that relate to the one THING. Confused? Don't be.

If you are at a loss for coming up with 4 THINGS OF THE THING, you can always take a short cut and write: "Never do this while playing golf" or "Always do this while playing golf."

Once you have mastered this concept, and some of you never will go beyond just the initial 16 THINGS which is okay, sit back and watching this move.

For each of the 4 THINGS OF THE THING, you will make 4 ADDITIONAL THINGS OF THE THING.

As an example, lets take the first THING OF THE THING, which is "golf in the morning or afternoon." Let's make 4 ADDITIONAL THINGS OF THE THING by speaking to the 4 personalities, which are:
1. The spontaneous reader.
2. The competitive reader.
3. The methodical reader.
4. The humanistic reader.

An example of the spontaneous reader is: (I want it now!)
"Play golf now as the course is empty."

An example of the competitive reader is: (what's in it for me)

"Be the first foursome out."

An example of the methodical reader is: (how can I do that)
"Step by step to finding the shortest waiting time."

An example of the humanistic reader is: (who is playing and what are their names)
"Who plays golf at 6 am."

So, from the one idea about golf, you have 16 THINGS, 4 THINGS OF THE THING and 4 ADDITIONAL THINGS OF THE THING. Let's look at the math.

16 things times 4 things of the thing equals 64 articles.

If you take 64 articles and create 4 personalities from each, you have 64 times 4 equals 256 articles. Wow, that wasn't too difficult. From one passion where you can write one article, you can create 256 articles. But wait...it gets better. Did you ever see those late night infomercials where they say if you order now, we will double your order?

How can anyone in his or her right mind sit down and write 256 articles without going insane? Easy.

If you need to write 256 articles between 400 and 600 words each, (let's use 500 words) simply break each article into 7 tips. Each tip is only 75 words. So, 75 words times 7 tips equals 525 words. Easy? Yes it is.
Make sure to design each article into 7 tips and fill in each tip with 75 words. Organize and set up each article with your 7 tips before you begin and it won't be a problem.

Once you do that, you can crank out 256 articles. Each article should take no longer than 30-minutes if you use this formula. That's a total of 256 articles times 30-minutes each equals 7,680 minutes. Divide 7,680 minutes by 60 (for one hour) and that's 128 total hours of work, or 3.2 weeks, less than one month.

The most cost effective way to approach this is hire a part time college student. Once each of the 256 articles has been identified and broken into 7 tips, so it will be easy for a writer or journalism major to expand each tip to 75 words. Your cost, for the 128 hours times $10 an hour would be $1,280 for about one months full time 40 hours a week working. Not bad. Make a flat offer of $1,000 to complete this assignment.

But wait...it gets even better. If you order in the next 30-minutes, you get 3 times the amount, plus process and handling. By the way, I am making fun of infomercials, but they really do draw the buyer in. So I am going to really draw you in on this one.

Somewhere along the line, I mentioned you can work once and get paid forever. If I didn't say that, I am saying it now. We call this "repurposing" your work.

So from one article idea, we got 256 articles. Now, I am going to show you how to:

Make 768 blogs.
Make 2,304 tweets.
Make 2,304 Facebook posts.
Make 2,304 LinkedIn posts.
Make 2,304 YouTube 2-minute videos.
Make 2,304 Blog Talk Radio Shows.
Make 2,304 Squiddo Lenses.
Make 256 free press releases.

Hopefully, by now you will see the power of repurposing and it will for sure makes sense to hire an additional part time college student who is social media competent.

Bonus of 2,304 Tweets

As promised, here is what happens when you follow the formula of Repurposing.

Now that you took one article and turned it into 256 articles, here is what you can do.

Each 500-word article can be further repurposed into 3 blogs. Blogs are shorter, more single focused summaries of a full-blown article. Simply chop up your 500 word article into 3 blogs, 150-200 words for each blog, slightly tweaked as a blog does not have to conform to E-zines editing rules and can be a more relaxed single message. It can be a sales pitch; can be a narrative or even a funny story. This is where you can have fun and let your hair down.

Having said that, here is the formula.

*Each article turns into 3 blogs. 256 articles times 3 equals 768 blogs.
*Each blog turns into 3 tweets. 768 blogs times 3 tweets equals 2,304 tweets.
*Each blog turns into 3 Facebook. 768 blogs times 3 Facebook posts equals 2,304 posts.
*Each blog turns into 3 LinkedIn. 768 blogs times 3 LinkedIn equals 2,304 posts.
*Each blog turns into 3 YouTube videos. 768 blogs times 3 YouTube videos equals 2,304 videos.
*Each blog turns into 3 Blog Talk Radio Shows. 768 blogs times 3 Blog Talk Radio Shows equals 2,304 radio shows.
*Each blog turns into 3 Squiddo Lenses. 768 blogs times 3 Squiddo Lenses equals 2,304 Lenses.
*Each blog turns into 3 Free Press Releases. 768 blogs times 1 Free Press Release equals 768 Free Press Releases.

Word Press and Blogger Blogs

Aim:
To use Word press and Blogger blogs as micro sales pages and as a free platform to drive traffic to your main site.

Domain Name Search:
It's all in the set up, meaning plan your entire brand out first before you begin so your domain name with one keyword will match your social media names and email. An example is:

www.janisspindel.wordpress.com
www.janisspindelmatchmaker.blogspot.com
www.twitter.com/janisspindel

www.facebook.com/Janis.spindel.matchmaker
www.blogtalkradio.com/matchmaker-Janis-Spindel

www.youtube.com/user/rockstar (I made a boo-boo with this name)
www.blog.janisspindelmatchmaker.com

Janis at janisspindelmatchmaker.com (the word "at" instead of "@" prevents spam)
www.janisspindelmatchmaker.com

Keyword Tool:
Use this free tool called Google Keyword Tool
http://www.googlekeywordtool.com/

Enter a keyword or keyword phrase.
Enter your web site.
Enter category.
Enter captcha and see what comes up.

It's self-explanatory and you can figure out what features are best for you. Download and export your customized list.

Spelling:
Many SEO consultants say make a list of keywords based on misspelled words such as match maker instead of matchmaker and plurals such as matchmaker matchmakers. Make a list of keywords that can be spelled more than one way such as Janis and Janice.

Pages:
Create pages on your Word press and Blogger blogs, as it's easy to create. Each new page can be its own unique web page to direct traffic for a specific marketing plan. http://janisspindel.wordpress.com/new-page/. Press PUBLISH when ready.

Pages must have the SEO Triple Play.
1. Keyword title.
2. Keyword description.
3. Keyword in body.

Left Side Bar:
In the settings section, you can change the title of the blog to confirm the keyword conforms to a triple play. I used Janis Spindel's Blog and then realized it should be Janis Spindel Matchmaking Blog to include the keyword matchmaking.

First time is the mistake. The title keyword does not match the description.

Second time is the corrected triple play. The title matches the description.

Posts:
Select New.
Keyword the title.
Keyword the description.
Keyword the first paragraph in the blog.
Link the keyword to one of your external blogs.

Triple play with blog title of matchmaking, post title of matchmaking, first sentence with keyword matchmaking and external link to a matchmaking blog.

All other buttons are self-explanatory and are critical, but don't want to get bogged down with that many Word press details in this eBook.

Press Publish when you are ready.

Paid Word Press Blogs:

In the Settings section, look at "All In One SEO" plug in.
You will see the Title Matchmaker.
Home Description is Matchmaker.
Home Keyword is Matchmaker.

The Word Press address (URL) http://askjanisspindel.com and the Site address (URL) http://www.askjanisspindel.com should have been with the keyword matchmaker in there. This was an oversight when I first created that blog. It should have been created "askmatchmakerjanisspindel.com."

"All in One SEO" plugin come with a paid Wordpress blog.

There are a number of other benefits of a paid Word press blog or paid Blogger blog that if you use www.godaddy.com, it will cost you about $11.50 a year to register and maintain a paid blog that would convert a free blog from www.janisspindelmatchmaker.blogspot.com to www.janisspindelmatchmaker.com as an example. Some other plug-in I found helpful with a paid Word press blog are RSS Feed and Outbound Click Tracker.

Now comes your first live exercise:
Create 5 blog titles using the Google Keyword Tool.
Create 5 blog sub titles from that same single keyword in the title.
Create 5 posts, making sure to use the same single keyword in the first paragraph used in the title and sub title.

This is called THE TRIPLE PLAY.

Summary:
Make sure your free or paid blogs have the Triple Play (title, subtitle and first paragraph) all have the same single keyword. Make sure each post is informative for the reader. Make sure the post is to the point so the reader walks away feeling really good about what they read.

Blog Talk Radio

Aim: Create 100 Blog Talk Radio shows with very strong SEO from Google, as Google will index that URL. Use these 3rd party authoritative URL's to market your web site and or brand using live-recorded radio shows that automatically create a Google URL.

Create 5 Blog Talk Radio shows at the same time each week for 5 weeks in a row. Each show will be 30 minutes and check what time zone you set it for.
Make sure you have the triple play with the Blog Talk Radio URL keyword http://www.blogtalkradio.com/matchmaker-janis-spindel matching the show information title keyword Janis Spindel Matchmaker that matches the description. America's top matchmaker Janis Spindel This is key.

Reminder, 3rd party authoritative links are the best food for Google to help your overall ranking.

Don't be disappointed if only a few active listeners call in, even if you social media and email your list with the URL and call in phone number.

The archive listener's column is a good indication of who searched you and who listens to the live recorded shows.

You can even document archived listeners and have someone sponsor the show. You would shout out 3X during the show. (Beginning, middle and end "And now a word from our sponsor") Give your sponsors 15-seconds of fame for each shutout.

Blog Talk Radio has a new premium service that will transcribe your live recoded shows. I have not as of yet used this service so I have no valid feedback to give you. Transcribing your shows allows you to repurpose that content on blogs and if you follow the formula in the beginning of this eBook, you can give each show long legs.

2 of the recorded shows have 55 and 59 Archive Listens and this is really good. From those listens, we all know that a certain percent will click on your link and purchase or subscribe to your product and or service. Make money while you sleep.

You can even submit your unique URL the DMOZ to insure Google can find it. We will talk about DMOZ in the next chapter.

Take that URL show and tweet it, post on Facebook and email to your list. Cross post on all your blogs using the triple play formula discussed in the beginning of this eBook.

Important note about recording your show. I did a few test shows using my cell phone sitting in the park under a tree and it came out really lousy. There was basically no sound and I could not hear the listeners nor could they hear me. I suggest using a landline with a good headset.

Conclusion: Blog Talk Radio is a great free viral marketing tool to drive targeted niche traffic to your web site and repurpose all content.

Free Press Release vs. Paid Press Release Sites

Aim: To build 3rd party authoritative links that editors and other influencers pick up. This allows your company profile to be exposed and possibly get picked up by Google News. You also want journalists and other bloggers to pick your press releases up so it appears high on a Google search.

Free press release sites exist mainly for themselves. By using your content and placing it on their sites in a uniform fashion, it actually makes their URL very strong in terms of Google ranking. If you think about it, they are nothing more than what is considered as a portal. A portal is a web site that brings information together from a collection of unique sources. In fact, each piece of information source has its own page. An example is http://www.i-newswire.com/janis-spindel-meets-online-matching/118999

The title is Janis Spindel Meets Online Matching.
The Body is just shy of pairing 1,000 married couples; Janis Spindel has built a reputation as a world-class matchmaker. Although this press release does not meet the perfect triple play, it was highly indexed by Google.

With in that unique URL and dedicated page, the free press release site runs Google Ad words all around it. They make money from your content; you get authoritative 3rd party links to place your content for others to view and also use, hopefully siting the original source. It's a big time win-win, providing your content is remarkable.

The negative about free press release sites vs. paid press release sites such as PR Web is many mainstream media companies don't pick up content from those free press release sites. In addition, Google doesn't index many free press release sites. However, my site www.janisspindelmatchmaker.com has been picked up and indexed by Google from the free press release site I-Newswire.

The rule of thumb is find 1 or 2 top ranked free press release sites and submit weekly articles. Many have restrictions and only allow a certain number of releases per month. My suggestion is check out a very well known and respected site called Mash able.

I have had a fairly good response rate and indexing by using I-Newswire. What I did find out after researching Mash able was that Online PR News, Open PR, PR

Fire, News Wire Today, PR Zoom Idea Marketers and I-Newswire came up favorably ranked.

Now what I really use and have had tremendous success with is a paid service called PR Web.

They charge an annual fee of $2,000 and give you 24 press releases for that year, which comes out to less than $85 per release. It's so worth it as PR Web is my number 1 referring site when I check Google Analytics.

Summary: Start out using 1 or 2 top ranked free press release sites, learn the ins and outs of writing press releases and monitor your traffic. If you get traffic to your site and you have a product or service to sell like I do, it makes economic sense to upgrade to a paid service.

If you search, notice how a Google search for "Janis Spindel press release sites," I-Newswire came up first, even before PR Web? This is very interesting.

That's okay because there is only one indexed URL from I-Newswire and there are other authoritative 3rd party links from a PR Web paid press release. Online Personals Watch picked up the paid press release. They are one of the most authoritative 3rd party sites in my industry. So in the long run, **paid press releases have "legs."**

DMOZ

Aim: In this chapter, you will learn how to submit your site to the DMOZ. The DMOZ is a free site where you list your new or current URL. It may take anywhere from 2 weeks to several months for your site to be indexed and listed on partner sites which use the open directory data, such as AOL Search, AltaVista, Hotbot, Google, Lycos, Netscape Search and more.

Select your category first and then select URL. Since I just built www.seolessonplans.com, I will actually do just that. My category is online teaching and learning.

Then look at the top navigation bar, the third to the right where it says Suggest URL. Click that and a yellow highlighted section will appear.

Follow the screen shots where you enter your site URL, title, and site description, email address, user verification and press the Submit button.

If done successfully, it a screen will appear saying submission received. Now comes the fun part.

I made a mistake and didn't have the triple play. I had Janis Spindel Blog instead of Janis Spindel Matchmaking Blog. Because my sites were all properly set up and indexed with Google, when I added that proper triple play, Google was kind enough to index me within 8 minutes.

It reads Janis Spindel Matchmaking Blog, www.janisspindel.wordpress.com and the 3rd line says 8 minutes ago. Thanks Google.

That's all you really need to do. My suggestion is test this theory. Create 2 free blogs using Word press or Blogger. One blog you submit to DMOZ and the other you don't. Wait a week and Google those blogs and track and document the results. Have fun and fine-tune your brands.

Conclusion: Listing your sites and blogs can greatly improve your organic search results for free.

Squiddo Lens

Aim: To build 100 Squiddo lenses to drive targeted niche free traffic to your site which helps increase the overall ranking of Google Searches for your industry keywords.

Squiddo is a free search tool that I came across in my research a while ago. The concept is very interesting and does help drive traffic to your site. It's so simple to set up, just follow the pre-defined fields.

Each Squiddo page can be a different micro message and is of different interest. Just tell your story, share an experience, or you can teach a course or sell a product. Each page is called Lenses, because as Squiddo says, it's a snapshot of your point of view.

Tell what your lens is about. See screen shot. Fill in the "my page is about" field and click continue.

Fill out each field such as the Lens Title, Set Your Lens URL, and Pick Your topic, Pick Your Subtitle, Pick One More Subtopic and Rate Your Lens. Click Continue.

Add 3 more Good Titles, enter the captcha and click continue.

You can edit your title, add your lens, intro title, add your lens description, even upload an intro image and away you go. Press Publish and you just created your own single page called a Lens.

Like all free Internet web based templates like Squiddo, you must realize that while they are helping you to build your business, so are they helping themselves through advertising. In fact, you can even specify if money is made from advertising, do you want to give it to a charity to do you want to receive revenue. Just fill out that section and away you go.

Conclusion: With just a few clicks of the mouse, you can create hundreds of free single-minded web pages to help drive great niche traffic to your main site.

Web Page Format

Aim: To properly set up your web site or blog so Google will rank you higher and maybe even on the first page of your industry keywords.

You don't need to know HTML code or any programming to build a high-ranking Word press or Blogger blog. All you need to know for these free blogs is the concept behind the "Heading Tags."

I will confess when I don't know something and at the time of the writing if this eBook, I know nothing about setting up heading tags. All that I need to know is Word press does it automatically.

All heading tags represent is the beginning of a new section of content on your blog. It's that simple.

So if your blog is set up properly with the triple play, (URL, title, sub title and first sentence of the first paragraph using the same keyword) then the search engine electronic spiders that crawl or check out your web site will have a very easy time finding it, which is a pleasant experience for them. As a result, they will reward your site and rank it higher than another site that made them work hard trying to find the information they were looking for.

That's all the technical stuff you and I need to know. The H1 heading is the most important and there might be 6 headings in total and the H6 is the least important.

For the purpose of this eBook, here is a very boring breakdown of heading tags and again, Word press does this for you when begin to type your blog and it will appear to others you are a genius when in fact, you might not be as I am for sure not that kind of genius.

```
<head>
<body>
<h1>Matchmaking Queen Does It Again</h1>
<h2>How The Best Matchmaking Service Married 1,000 Couples</h2>
<h3>Americas top matchmaking service launches first event with 150 upscale
Asian-Americans in attendance</h3>
</body>
</head>
```

Now, don't take this HTML code to the bank because I copied it from one of my 24 blogs and its possible I messed it up when typing, but you get the idea. The lower the heading tags, the more towards the body of the blog you get. Now comes the good stuff.

It's said that writing content for the web is very different than being an editor at a newspaper and its true. Web visitors are much different because they are searching for more information at the speed of light and you have 5-15 seconds for them to decide if they want to continue reading.

Therefore, your web site or blog must be designed to appeal to online visitors who land on it or you will loose them forever. So here is what I have learned and you decide for yourself if you buy into it or not. I have personally tested this theory and my hypothesis was correct.

Question: How do YOU know what color background your web site should be?
Answer: YOU don't know what color visitors will prefer.
Solution: Build 2 identical pages with 2 different color backgrounds.

Now you are for sure a genius. You just solved the problem of knowing which background color YOUR visitors prefer. Do you see where I am going with this?

The only real way to know for sure is what the industry calls "split testing or A/B testing. The A page is color blue and the B page is color green as an example. Wait 2 weeks and check to see how long visitors stay on each test page using Google Analytics, or how many call to actions each site generates in a 2 week period. Once you determine which page produces the best results that color wins. Repeat the test taking the winner and make that page A and test with another color and make that page B.

Wait another 2 weeks and take the winner and test it against yet another color. Repeat this 5-10 times and then you are guaranteed to know which color YOUR visitors like.

That's the good news meaning you found out what is the best background color for your market. Here comes the bad news. You guessed it. You need to test several other components one at a time. The other components to test are:

Font color, font size, font style, layout, graphics, icons, headings, the copy, the submit button, the use of images, the use of social media icons and free give a ways.

1. Components of the upper fold. That is the display before you scroll down. Split test the banner, the logo design, tabs, images shown at the top and headlines.

2. Presentation of the product or service you offer. Display a product or service with a winning snap shot that best represents the product sold.

3. Call to action. After you make the presentation of your products or service in a catalogue style, you need to leave a call to action. This is best displayed with a snippet of text usually located at the end of your product description or other information that convinces the visitors to take your action such as "buy now."

4. Layout. This suggests that web visitors follow a certain pattern when surfing the web and viewing web pages. The movement of the eye can go from up the page to down the pages or from left to right. Researchers have concluded that eyeballs begin at the right side of the page and move left.

That's basically it. The more time you have, the more components you can test and make sure to test only one component at a time.

I built 2 blogs to test if viewers like to read blogs or watch a video. The first blog text to read and a side bar video. The second blog has the video smack in your face for viewers who like to watch videos first.

Conclusion: The more components you test, the more time you take to tweak your web site pages, the better results you will get from conversions and calls to action.

Proprietary Blog O Sphere

Aim: To build and dominate your industry through a group of proprietary blogs based around your keywords all driving targeted organic niche traffic to your main site to sell products and services.

In this final and most critical chapter you will learn how to put all the previous chapters knowledge together to dominate your industry keywords.

Currently, I have 24 blogs all in the matchmaking and dating industry with keywords including Janis Spindel that point to the mother site www.janisspindelmatchmaker.com. Recently, a young woman approached Janis Spindel about starting an Asian-American matchmaking division and I experimented and learned something very interesting.

Once you create your first web site or blog and have determined the layout by split testing that becomes the main site. One of the key concepts is adding the HTML code "powered by" and add your URL at the bottom of your site. Here is the HTML my web master set up and just replace your URL with mine.

<a href="http://www.janisspindelmatchmaker.com/" target="_blank">Powered By Janis Spindel Serious Matchmaking, Inc.</a>

Google seems to love this html snippet and when you place that HTML code in all your blogs, Google counts all the traffic from all your blogs collectively and that's very powerful for overall search engine ranking.

Now, create 5 URL's, 5 titles, 5 sub titles and 5 first sentence blog posts. Each blog triple play has the same keyword and each blog has a different keyword. The end result is you have 5 blogs with 5 different keywords.

As an example, this is my matchmaking business proprietary blog o sphere.

<a href="http://www.janisspindelmatchmaker.com">Click here</a> to begin your Journey!!!

<a href="http://www.asiamatchmaking.blogspot.com">Asia Matchmaking</a>

<a href="http://www.easianlove.blogspot.com">eAsian Love</a>

```html
<a href="http://www.dianekimmatchmaker.blogspot.com">Diane Kim
Matchmaker</a>

<a href="http://www.asianamericanmatchmaking.blogspot.com">Asian American
Matchmaking</a>

<a href="http://www.koreanamericanmatchmaking.blogspot.com">Korean
American Matchmaking</a>

<a href="http://www.matchmakerkorea.blogspot.com">Matchmaker Korea</a>

<a href="http://www.matchmakerasia.blogspot.com">Matchmaker Asia</a>

<a href="http://www.ekoreanlove.blogspot.com">eKorean Love</a>

<a href="http://www.asianmatchmaker.blogspot.com">Asian Matchmaker</a>

<a href="http://www.lovewithdiane.blogspot.com">Love With Diane</a>

<a href="http://www.koreanmatchmaking.blogspot.com">Korean Matchmaking
</a>

<a href="http://www.asianlovematch.blogspot.com">Asian Love Match</a>

<a href="http://www.koreanmarriage.blogspot.com">Korean Marriage</a>

<a href="http://blog.janisspindelmatchmaker.com/">Janis Spindel Blog</a>

<a href="http://blog.janisspindel.wordpress.com/">Janis Spindel Word Press
Blog</a>

<a href="http://www.datingdivausa.com/">Dating Diva USA</a>

<a href="http://www.janisspindeluniversity.com/">Janis Spindel University Digital
Downloads</a>

<a href="http://www.askjanisspindel.com/">ASK Janis Spindel</a>

<a href="http://www.janisspindelsvipmensvirtuallounge.com/">Janis Spindel VIP
Men's Lounge</a>
```

```
<a href="http://www.vipmensvirtuallounge.com/">VIP Men's Lounge</a>
<a href="http://www.matchmakerspindel.blogspot.com">Matchmaker Spindel</a>

<a
href="http://www.matchmakerslogicallocations2pickupmen.blogspot.com">Match
makers Logical Locations 2 Pick Up Men</a>

<a href="http://www.datingtipsprocess.blogspot.com">Dating Tips Process</a>

<a href="http://www.janisspindelmatchmaker.blogspot.com">Janis Spindel
Matchmaker </a>

<a href="http://www.datingadviceprocess.blogspot.com">Dating Advice
Process</a>
```

Conclusion: Create your own blog o sphere and make sure all blogs are linked and cross-linked. I can assure you that if you follow my step-by-step process; you will see dramatic improvement in your search engine ranking.

www.ingramcontent.com/pod-product-compliance
Lightning Source LLC
Chambersburg PA
CBHW081248170526
45165CB00009B/3243